Your World

Shoes

Classifying

Dona Herweck Rice

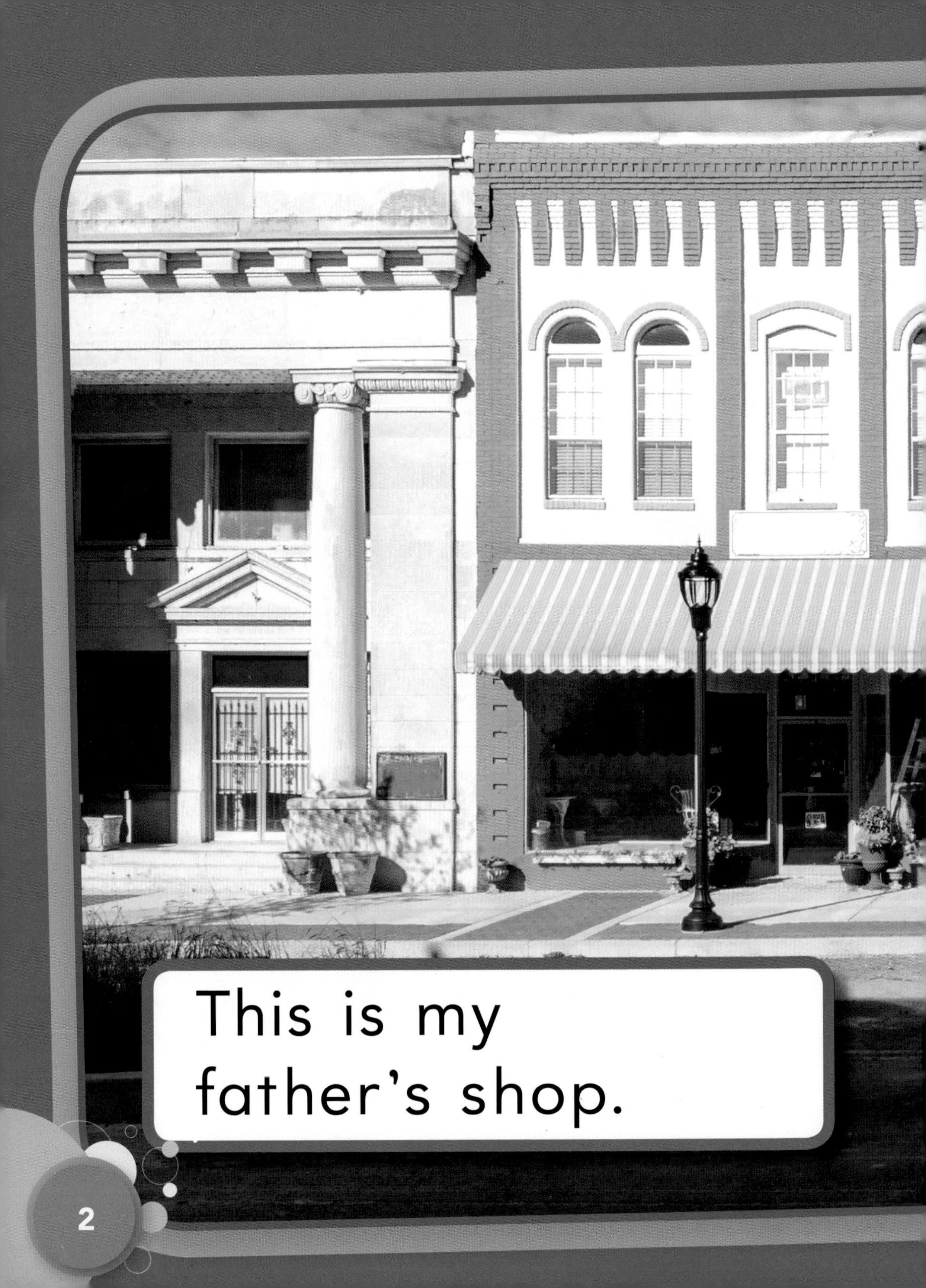

This is my father's shop.

I like to help him.

First, I get
the boots.

I sort them
by type.

Next, I get
the sandals.

I sort them by
number of straps.

Next, I get the sneakers.

I sort them by size.

Next, I get
the loafers.

I sort them
by color.

Last, I get the
high heels.

I sort them by heel height.

Uh-oh!

I forgot to sort
the shoes in pairs!

Problem Solving

My father sells socks at his shop, too. Sort the socks into two groups.

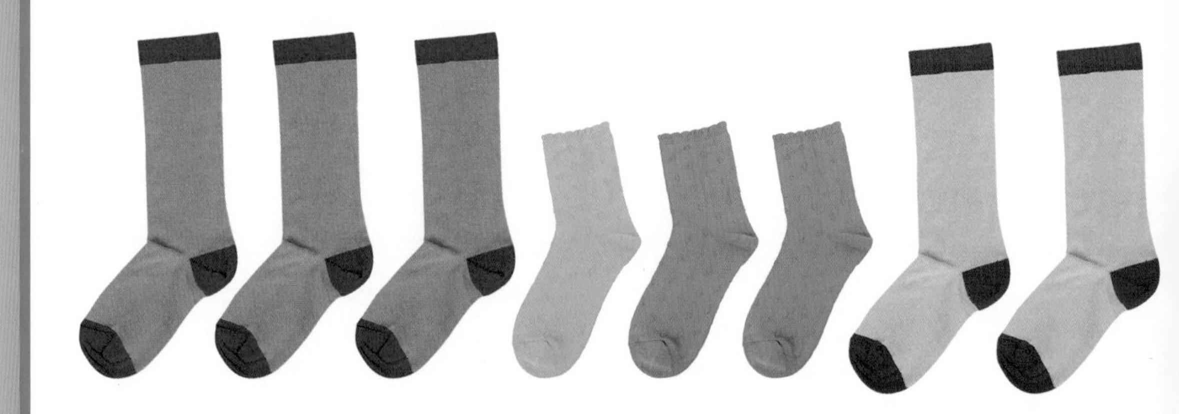

1. Draw a picture to show your groups.

2. How many socks are in each group?

3. Compare your groups using *most*, *least*, or *the same as*.

4. Draw a picture to show another way to sort the socks.

Answer Key

1. Pictures should show two groups of socks sorted into categories.
2. Answers will vary.
3. Answers will vary.
4. Pictures will vary but may show sorting by size, color, or design.

Consultants

Nicole Belasco, M.Ed.
Kindergarten Teacher, Colonial School District

Colleen Pollitt, M.A.Ed.
Math Support Teacher, Howard County Public Schools

Publishing Credits

Rachelle Cracchiolo, M.S.Ed., *Publisher*
Conni Medina, M.A.Ed., *Managing Editor*
Dona Herweck Rice, *Series Developer*
Emily R. Smith, M.A.Ed., *Series Developer*
Diana Kenney, M.A.Ed., NBCT, *Content Director*
June Kikuchi, *Content Director*
Véronique Bos, *Creative Director*
Robin Erickson, *Art Director*
Stacy Monsman, M.A., and Karen Malaska, M.Ed., *Editors*
Michelle Jovin, M.A., *Associate Editor*
Fabiola Sepulveda, *Graphic Designer*

Image Credits: pp.18–19 EQRoy/Shutterstock; all other images from iStock and/or Shutterstock.

Library of Congress Cataloging-in-Publication Data

Names: Rice, Dona Herweck, author.
Title: Your world : shoes / Dona Herweck Rice.
Description: Huntington Beach, CA : Teacher Created Materials, [2019]
Identifiers: LCCN 2017059894 (print) | LCCN 2018002579 (ebook) | ISBN 9781480759541 (e-book) | ISBN 9781425856168 (pbk.)
Subjects: LCSH: Shoes--Juvenile literature. | Classification--Juvenile literature.
Classification: LCC GT2130 (ebook) | LCC GT2130 .R53 2019 (print) | DDC 391.4/13--dc23
LC record available at https://lccn.loc.gov/2017059894

Teacher Created Materials
5301 Oceanus Drive
Huntington Beach, CA 92649-1030
www.tcmpub.com

ISBN 978-1-4258-5616-8

Printed in China
Nordica.072018.CA21800711